ARE THE LAWS
OF PHYSICS
WEIRD?

JOSEPH **PALAZZO**

authorHOUSE°

AuthorHouse™
1663 Liberty Drive
Bloomington, IN 47403
www.authorhouse.com
Phone: 1 (800) 839-8640

Published by AuthorHouse   06/29/2016

ISBN: 978-1-5246-0531-5 (sc)
ISBN: 978-1-5246-0530-8 (e)

Print information available on the last page.

This book is printed on acid-free paper.

THIS BOOK IS DEDICATED

TO

RITA PALAZZO

CONTENTS

It was Nima Arkani-Hamed[1] who once said, "...**take the laws of physics and push them to their absolute limits**..." An alternative is to go back to square one and re-examine our fundamental assumptions. It is that logic that led me to uncover the assumption underlying Galileo's Law of Inertia, an assumption that infiltrated undetected into Einstein's theory of General Relativity (chapter 1). The same logic led me to uncover four major misunderstandings that are at the base of the Einstein-Bohr disagreement on the interpretation of Quantum Mechanics, a long history mired into paradoxes, misconstructions and confusion which culminated into Bell's theorem, itself deeply misunderstood (chapter 2). Lastly, chapter 3 is about a toy model of a universe of bouncing balls that helped me to discover a new law of kinematics. The logic behind that new law brings new insights into Thermodynamics, Relativity and Quantum Field theory.

Part of this book is to dispel the notion that the laws of physics are weird. The other part is to present them as basically logical. I can only write this with humility and gratefulness that I arrived late in my life at this junction when logic spurred on this creativity in my mind. I can only hope it will help all those who are serious about solving our problems on a rational basis, and not let feelings override rational thoughts. As humans, we are all prone to fall into the trap of irrational fears and pervasive hate that beset so many in a world awash with disinformation. My sincere wish is that this book will instill in you a new hope.

Joseph Palazzo

Chapter 1

A New Interpretation of General Relativity

Preliminary: Galileo's Law of Inertia is corrected, and Einstein's Theory of General Relativity (GR) is an extension of that correction.

Einstein arrived as his Equivalence Principle[2] (EP) with the following thought experiment. In a room without windows, one would not be able to distinguish between being pulled up by an inertial force F_I (Fig. 1a) from being pull down by a gravitational force F_G near a planet (Fig. 1b).

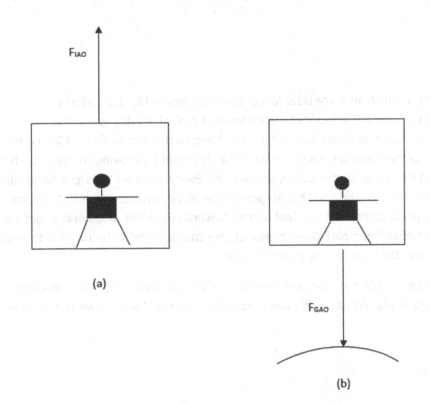

(a)

(b)

Fig. 1.1

This is not entirely true. There are tidal forces that would show differences.

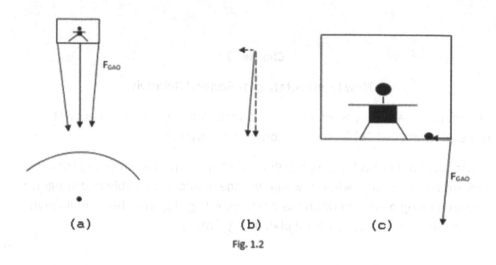

(a)   (b)   (c)

Fig. 1.2

Acting on the frame are tidal forces as illustrated in Fig. 1.2. What the observer would feel besides the downward pull of gravity, is a force perpendicular as illustrated in fig 1.2b acting to the left, and Fig. 1.2c, where a ball is placed near a corner, showing that it would be moving to the left. There would be similar forces acting to the right. Every point on the right-hand side in that frame would feel this perpendicular force towards the left. Similarly, every point on the left will feel a force towards the right. But what about the Equivalence Principle? Does it predict also this perpendicular force in the case of an inertial force acting on the frame?

To examine this question, we need to realize that an inertial force does not exist ex nihilo. What could have created this inertial force as we see it in Fig. 1.1a?

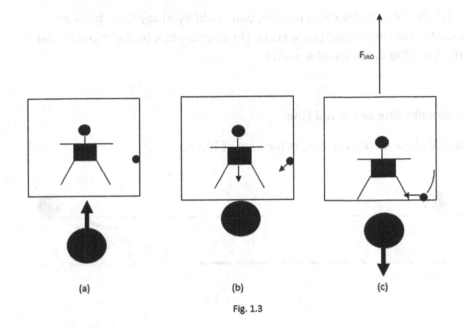

(a)   (b)   (c)

Fig. 1.3

Consider a collision, Fig. 1.3. Initially you and the ball are weightless, floating in space, Fig. 1.3a.

On the basis of Newton's 3rd law of motion, the frame applies a force on the colliding body, while the colliding body applies on the frame a force of equal magnitude but opposite direction. You will be pulled towards the bottom of the frame (Fig. 1.3b). We also see that after the collision, the floating ball in your frame experiencing a force that has a perpendicular component, will hit the bottom and accelerate towards the left (Fig. 1.3c). Now, since you are in a room without windows, you could say that:

(a) Either an inertial force acted on you as in Fig 1.1a,
(b) Or you have enter a gravitational field as in Fig. 1.1b.

In both cases you would see the ball accelerating towards the center of the room.

In either case, you could conclude:

(1) Either there are perpendicular (tidal) forces acting on the ball,

(2) Or if you know some physics, you could even say that this is an illustration of space-time being curved in a region due to the gravitational force from the presence of a planet.

1.1 Straight Line or Curved Line

Consider how Galileo arrived at the Law of Inertia.

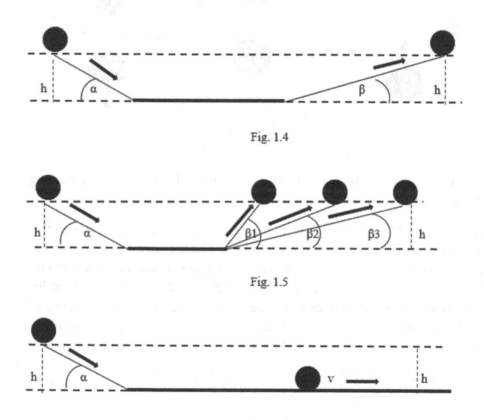

Fig. 1.4

Fig. 1.5

Fig. 1.6

A ball is released from a height h on an inclined plane making an angle α with the ground (fig. 1.4). It would then roll down and climb a second inclined plane making an angle β with the ground. What Galileo observed is that the ball would travel far enough until it would climb a height of h where it would come to rest temporarily and then reverse course. By varying the angle of the second plane (Fig. 1.5), he always observed that the ball would climb to a height h before reversing direction.

This led him to a thought experiment: what would happen if the second inclined plane were to be removed (Fig. 1.6)?

He reasoned that the ball would continue to move endlessly in a straight line with a velocity v trying to reach the height h, unless other forces would compel it to change its velocity. Now this falls into the realm of an assumption as it cannot be proven in the real world - there are always other forces that will act on the ball, for instance, friction to name one.

But is the ball really going in a straight line?

Had Galileo extended his experiment on a much larger lengthy floor, he would have realized that his ball was actually moving along an arc, called a geodesic, as in Fig. 1.7.

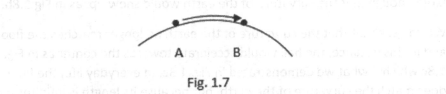

A   B

Fig. 1.7

We can now restate the Law of Inertia: unless other forces compel it to change its velocity, an object will continue endlessly with a velocity v, along a curved line. The alternative to that, which would be the modern version, is the ball entered a region of curved space. Fortunately, Galileo escaped this conclusion, otherwise his law would have been met with great skepticism.

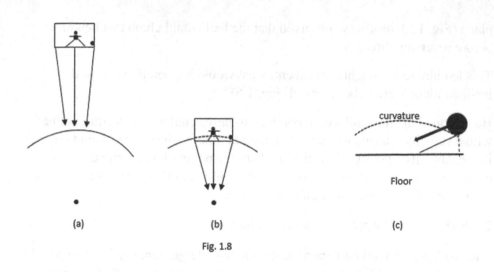

curvature

Floor

(a)   (b)   (c)

Fig. 1.8

Suppose in Fig 1.2a, reproduced above in Fig. 1.8a, the room is sufficiently large enough that the curvature of the earth would show up, as in Fig 1.8b.

We can see that that the curvature of the earth no longer matches the floor, and in this instance, the ball would accelerate towards the center as in Fig. 1.8c which is what we demonstrated in Fig. 1.3c. In everyday life, the floor does match the curvature of the earth, but because its length is infinitesimal small compared to the curvature of the earth, we are led to think as in the original Galileo's experiment, believing that the ball would continue to move with velocity v along a "straight line".

When Newton proposed his universal Law of Gravity it was met with great skepticism as it was after all a spooky action at a distance. And Einstein often repeated that his GR had explained away this spooky action at a distance. Nevertheless, his theory replaced "spooky action at a distance" with "space-time is curved". One can debate over which one is the least palatable.

1.2 Intensity or Force

Consider a candle radiating light in all direction (Fig. 1.9a).

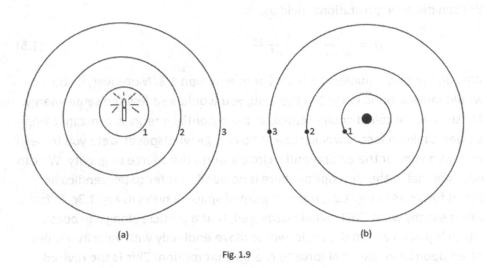

(a)   (b)

Fig. 1.9

The energy released by the candle spreads out in concentric circles. Since energy is conserved, we find that at point 1 in Fig 1.9a,

$$I_1 = \frac{k}{4\pi R_1^2} \tag{1.1}$$

Where k is the energy released at the origin by the candle, and $4\pi R_1^2$ is the surface area of the sphere over which the energy has spread out at R_1.

Similarly for points 2 and 3,

$$I_2 = \frac{k}{4\pi R_2^2} \quad and \quad I_3 = \frac{k}{4\pi R_3^2} \tag{1.2}$$

In general, for any distance R:

$$I = \frac{k}{4\pi R^2} \tag{1.3}$$

If you would move around circle 1, you would feel the same intensity (energy). Ditto for circles 2 and 3. We have a similar situation with gravity in Fig. 1.9b. Consider a point mass instead of a candle that serves as a source of gravity, and a test mass that we can place on any of the circles 1, 2 or 3.

According to Newton's Law of Gravity,

$$F_g = \frac{GM_{source}M_{test}}{R^2}$$

(1.4)

We can define a gravitational field as,

$$g = \frac{F_g}{M_{test}} = \frac{GM_{source}}{R^2}$$

(1.5)

We can see that equation 1.5 is similar to equation 1.3. Moreover, if you would move around circle 1 in Fig. 1.9b, you would also feel the same energy. These lines are called equipotential, as the potential energy is constant along a given circle. In fact, it would require no energy whatsoever were you to be in motion on any of the equipotential circle around the source of gravity. We can now see that in this description there is no need to refer to perpendicular (tidal) forces as in Fig. 1.2c, nor to a warped space-time as in Fig. 1.3c. In fact this is exactly what Galileo had discovered, that a particle along a geodesic, which is just an arc on the circle, would move endlessly with velocity v unless acted upon by an external force to change that motion. **This is the revised Law of Inertia.**

It was no coincidence that $M_{initial} = M_{gravitational}$. In fact, it was unnecessary to distinguish an inertial mass from a gravitational mass, as they are not only equivalent, but also they are exactly the same. In the revised Law of Inertia, the mass in Galileo's Inertia Law is a gravitational mass.

1.3 Do We Live in a 4-D World or a 3-D World?

Consider light and its speed c,

$$c = \frac{dr}{dt}$$

(1.6)

Where in a 3-D world,

$$dr = \left(\sum_{i=1}^{3} dx_i\right)^{1/2}$$

(1.7)

Squaring both equations 1.6 and 1.7,

$$c^2 = \frac{dx_i dx^i}{dt^2}$$

(1.8)

14

Where we have used Einstein summation rules. Multiply the two sides by dt^2,

$$c^2 dt^2 = dx_i \, dx^i \qquad (1.9)$$

Rearranging,

$$- c^2 dt^2 + dx_i \, dx^i = 0 \qquad (1.10)$$

Since the speed of light is a constant in every reference frame, then equation 1.10 is true in every frame of reference.

The 4-vector Formalism

Define a 4- vector $x_\mu = (x_0 , x_i)$ where Latin index = 1,2,3 ; and Greek index $\mu = 0,1,2,3$, and $x_0 = ct$.

Define a 4 × 4 matrix, $\eta_{\mu\nu} = \begin{pmatrix} -1 & 0 & 0 & 0 \\ 0 & 1 & 0 & 0 \\ 0 & 0 & 1 & 0 \\ 0 & 0 & 0 & 1 \end{pmatrix}$ called the metric tensor.

Now we can write equation 1.10 as,

$$\eta_{\mu\nu} \, dx_\mu dx^\nu = 0 \qquad (1.11)$$

Notice we started out in equation 1.6 with a 3-D equation. That we define a 4-vector formalism in equation 1.11 does not change the universe from a 3-D reality to a 4-D reality. The 4-vector formalism is convenient for our needs – it makes the equation more compact, more esthetically pleasing, and easier to manipulate algebraically, but let's not fool ourselves. It would be presumptuous to think that the universe should bow down to our whims!

For a massive particle, we write equation 1.11 as,

$$ds^2 = \eta_{\mu\nu} \, dx_\mu dx^\nu \qquad (1.12)$$

For a particle moving along a straight line in the x-direction

$$ds^2 = -c^2 dt^2 + dx^2 < 0 \qquad (1.13)$$

That's true because no massive particle can travel faster than light and so $dx < cdt$. However we need to think of the term "cdt" as the distance travelled by light, and not as a time coordinate.

1.4 From Galilean transformation to Lorentz Transformation

First law of Kinematics

Frame 1   Frame 2

Fig. 1.10

Galileo's inertial law of motion, ignoring the curvature of the earth for this case, can be represented as one body viewed in two different frames of reference.

We have two observers: one is at rest with the body, Frame 1; the other observer is in uniform motion with respect to the same body in question, Frame 2.

Second Law of Kinematics: The Law of Transformation

This law is concerned with how in different inertial frames of reference such as Frame 1 and 2 the laws of nature are transformed. Initially, we had the Galilean transformation equations.

$$x_2 \rightarrow x_1 - vt \qquad (1.14)$$

In the theory of Special Relativity (SR) Einstein made us all aware that in order to have the constancy of the speed of light in every inertial frame of reference, the Galilean transformation equations had to be replaced by the Lorentz transformation equations.

$$x_2 \rightarrow \gamma(x_1 - vt) \tag{1.15}$$

$$t_2 \rightarrow \gamma(t_1 - vx_1/c^2) \tag{1.16}$$

Where $\gamma = \left(1 - \dfrac{v^2}{c^2}\right)^{-\frac{1}{2}}$

The surprising result was that time is no longer an absolute quantity but when measured in a given frame it will yield a different time measurement compared to another frame, and this effect is generally known as time dilation.

The argument that the proponents of a 4-D universe is that the time t gets mixed up with the position x in the Lorentz transformation equations. And therefore it acts as another coordinate. However, one should think of equation 1.16 as,

$$ct_2 \rightarrow \gamma\left(ct_1 - \frac{v}{c}x_1\right) \tag{1.17}$$

It's just that one shouldn't forget that not only do we get time dilation, but also length contraction, so our length ct_2 as measured in the second frame will be shortened.

1.5 Fields or Energy Levels

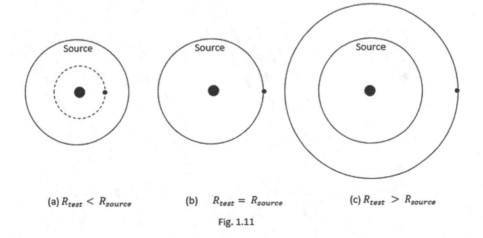

(a) $R_{test} < R_{source}$ (b) $R_{test} = R_{source}$ (c) $R_{test} > R_{source}$

Fig. 1.11

As it was mentioned above, around a spherical object, we have equipotential energy levels. Also, we have a source and a test particle in a field. It is understood that the test particle is taken to be so small that its own field compared to the source field is negligible.

In Fig. 1.11, we have three cases in which,

(i) The test mass is placed inside the source, $R_{test} < R_{source}$ (Fig. 1.11a).

(ii) The test mass is placed on the surface of the source, $R_{test} = R_{source}$ (Fig. 1.11b).

(iii) The test mass is placed outside the source, $R_{test} > R_{source}$ (Fig. 1.11c).

In all cases, the test mass will experience a force of gravity as if all the mass inside the circle on which it resides is concentrated at the center. In case (i), the test mass will experience only the force due to the mass inside the dotted line, in other words, a reduced source mass.

The good news is that we can ignore everything outside the circle of equipotential energy on which the test mass resides, even if the rest is a universe of infinite in mass (Fig. 1.12). For the test particle, the only thing that matters is the mass inside the sphere defined by the circle of equipotential energy. We can literally say that gravity lives on a sphere.

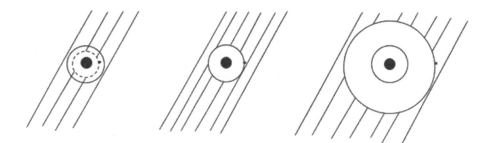

(a) $R_{test} < R_{source}$ (b) $R_{test} = R_{source}$ (c) $R_{test} > R_{source}$

Fig. 1.12

1.6 Einstein's "Field" Equation

Of course if you insist on working your physics on a local "inertial" flat small patch[2], "inertial" as defined originally by Galileo along a straight line rather than a geodesic, you will experience those weird effects – "there are tidal forces" or "space-time is warped" – as such a patch does not match the curvature of the earth (a spherical object) as demonstrated in Fig. 1.7. In our everyday life, we do not experience these weird effects simply because when we walk around, we are moving along the surface of the earth, an equipotential surface. When we go up an inclined plane, we must exert ourselves to overcome the force of gravity, or when we go down, which it's an easier task as gravity is pulling us down. But we still do not experience these weird effects as our inclined plane is not a flat small patch but rather a small curved surface. Since gravity lives on a sphere, any other kind of coordinate system, especially the flat ones, will produce these weird effects.

Are these effects real? They are as real as the Coriolis force is. Pilots must take into account of the Coriolis force when flying their planes to destination; similarly tidal forces explain the high and low tides of the oceans. But no one dreams of quantizing the Coriolis force. We'll leave to others to decide if quantizing gravity is a worthwhile endeavor.

But now let us examine closely what are the Einstein's "field" equations and what do they really mean.

Einstein reasoned that if "space-time" is curved, then equation 1.12 must be amended,

$$\eta_{\mu\nu} \; \rightarrow \; g_{\mu\nu} \; and \; ds^2 = g_{\mu\nu}dx_\mu dx^\nu \qquad (1.18)$$

Then one can use SR to identify[3][4],

$$g_{00} \; \rightarrow \; -\left(1 + \frac{2\varphi}{c^2}\right) \qquad (1.19)$$

Where φ is the gravitational energy potential, and not the field, which is:

$$\varphi = -\frac{GM_{source}}{R} \qquad (1.20)$$

It's not a coincidence that the first solution proposed by Schwarzschild for a massive "spherically-symmetric" object yields the Schwarzschild radius as,

$$R_{\text{Schwarzschild}} = \frac{2GM}{c^2} \tag{1.21}$$

Then from Newton's law expressed in terms of Gauss' theorem,

$$\nabla^2\varphi = 4\pi\rho \tag{1.22}$$

Where ρ is the mass density, Einstein identifies,

$$T_{00} \rightarrow \rho \tag{1.23}$$

Where T_{00} is the energy-momentum tensor. The rest is history. The Einstein "Field" equations are then,

$$R_{\mu\nu} - \tfrac{1}{2}g_{\mu\nu} = \frac{8\pi G}{c^4}T_{\mu\nu} \tag{1.24}$$

Disregarding the factor in front of the energy-momentum tensor, we see that the Einstein's "Field" equations have a lot more to do with energy than a field. It stands to reason that objects want to follow the path of a geodesic, because that's where equipotential circles exist. It's where planets establish their stable orbits, it's what we as humans do when we walk around effortless on the surface of the earth, and it's where a gravitational mass will move endless with velocity v unless an external force compels it to do otherwise. If we insists on describing the laws of physics on a flat coordinate system moving along a "straight" line, which doesn't match a geodesic, then the weird stuff comes out.

Chapter 2

The Collapse of the Wave Function

2.1 The Heisenberg Uncertainty Principle (HUP) Revisited

A re-interpretation of the HUP is in order.

Here's a thought experiment. Suppose you were God and you could grab an electron and deposit at a certain position. As God, you've just violated the HUP – but that's okay, God can do that. We could depict this as in Fig. 2.1.

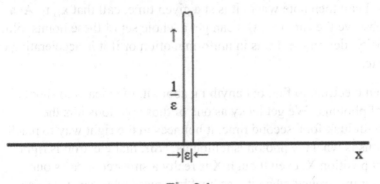

Fig. 2.1

But as soon as you release the electron, it would look like Fig. 2.2.

After a time T, the position of the electron has spread out. The question is: what does that tell us? It looks like the particle is doing some kind of motion, some jiggling. It means that for microscopic particles, they are never at rest. In classical physics, you can have objects at rest. The walls in your room are at rest with you. But in QM, no object is at rest. And that's a fundamental difference with classical physics.

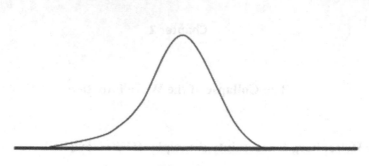

Fig 2.2.

What else does the HUP tells us?

How do we measure the velocity of a car? I see the car because an enormous number of photons are hitting the car in all directions, and some of them will reach my eyes. I can then note where it is at a given time, call that x_1, t_1. At a later times, I observe the car at x_2, t_2. I can get a whole set of these points, plot it, get the velocity, determine if it is in uniform motion or if it is accelerating or decelerating, etc.

So it goes for an electron, to find out anything about it, the idea is to shoot a whole bunch of photons. We get lucky as one of those photons hits the electron, and with luck for a second time, it bounces in the right way to reach our eyes. But this is what the photon is telling us, "Sir, that electron is right there," call that position X, even though X is really a smeared area as our electron was jiggling around when it was hit, "but guess what Sir, I've also thrown it off its position, and I haven't a clue in what direction it's going." This is the second thing the HUP is saying: if the position is known with zero uncertainty, then its momentum is unknown. And likewise, if the momentum would be known with absolute certainty, then its position would also be unknown. This is characterized as,

$$\Delta\sigma_x \, \Delta\sigma_p \; \geq \; \hbar/2 \tag{2.1}$$

Where σ is the standard deviation. Note that in the case of the car (a classical system), we need not to worry that the photons will disturb the trajectory of the car. We will explore more of this meaning later.

The third thing that the HUP says is that if you make a measurement, the very act of making the measurement will alter the system. In our case, we had an electron jiggling about the position X, and now, it's jiggling somewhere else.

To resume, for a quantum system:

(1) A particle is never at rest.
(2) There is an uncertainty in measuring the position and momentum at a given time as indicated by equation 2.1.
(3) A measurement on the quantum system alters the system in some unpredictable way.

The net result is that we get partial knowledge of the quantum system, and we have to make do with that reality.

2.2 Incompatible Observables – Conjugate Pairs

Consider a number of plane waves moving towards a slit as in Fig. 2.3.

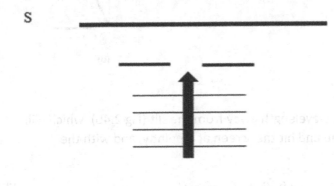

Fig 2.3

As they go through the open slit, they start to bend and will hit the screen S, leaving on it a series of white and dark fringes (Fig. 2.4a).

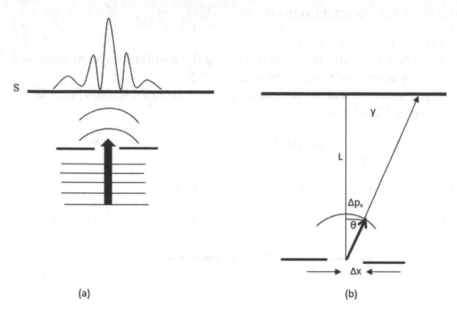

S

(a)   (b)

Fig. 2.4

Consider a point one wavelength away from the slit (Fig 2.4b), which will travel in a straight line and hit the screen at a point y, and with the wavelength $\lambda \ll L$.

$$sin\ \theta \approx \theta = y/L \qquad (2.2)$$

Also, when that point was entering the slit, it had only a momentum p along L. But now it's moving at an angle θ and has developed a momentum along the x-direction, Δp_x.

$$\Delta p_x \approx p\ \theta \qquad (2.3)$$

Also, the point on the wave is one wavelength λ, and $\Delta x/2$ away from the line of motion,

$$\lambda \approx (\Delta x/2)\ \theta \qquad (2.4)$$

Now, historically, de Broglie had proposed that every particle is associated with a wavelength such that,

$$p = h/\lambda \qquad (2.5)$$

Where h is Planck's constant. Combining 3, 4 and 5, we get

$$\Delta x \, \Delta p_x \approx 2h > h \qquad (2.6)$$

This is also the HUP, expressed in terms of the uncertainty in the position Δx, and the uncertainty in the momentum Δp. If we want to reduce the uncertainty in the position by passing the waves through a smaller slit, then the bending of the waves will be more pronounced, and so the uncertainty in the momentum will be larger. And vice versa, to decrease the uncertainty in the momentum requires less bending, and to accomplish that the slit must be wider. We say that the position and the momentum are incompatible observables. These are often called conjugate pairs.

2.3 Quantum States

A lot of confusion in Quantum Mechanics is the result from not being able to differentiate between the real world and the Hilbert Space. Vectors in real space – like velocities, accelerations, forces, etc. – are objects one can actually measure in the real world. On the other hand, quantum states are represented by vectors (more precisely by rays) in a Hilbert space, but these are NOT subjects of measurement. What we measure for a quantum system are probabilities, and those vectors in that Hilbert space are useful mathematical tools to calculate those probabilities.

Suppose we have a beam of electrons flowing from right to left (Fig. 2.5):

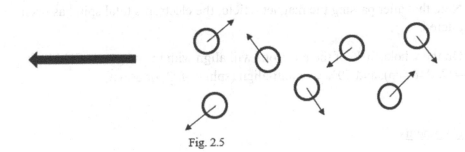

Fig. 2.5

Notice this is a thought experiment as we really don't know in what direction the spin of each individual electron points. We can safely say that these directions are at random. Now we are interested in measuring these spins. So what is needed is some kind of apparatus, and the good news is that there exists one – a magnetic field. Trouble is that these electrons, with their spin, are tiny magnets, and we know that magnets placed in a magnetic field will align (or anti-align) with the magnetic field. Suppose a magnetic field is placed along a certain direction, say the z-axis. Now let's look at one electron as it approaches the magnetic field (Fig. 2.6c).

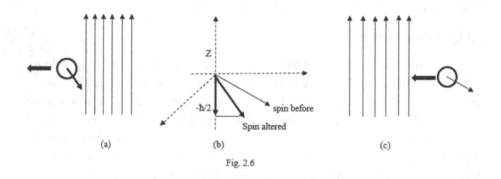

Fig. 2.6

When that electron penetrates the magnetic field, it will align its spin such that its z-component will yield the value of $-\hbar/2$ along the z-axis, a spin down, which can be represented as in Fig 2.6b.

Note that after passing the magnetic field, the electron's total spin has been altered.

On the whole, 50% of the electrons will align with the magnetic field (spin $=+\hbar/2$, or up), and 50% will anti-align (spin $= -\hbar/2$, or down).

Comments

(i) Before the measurement, the spin of an electron can be in any direction. Passing the electron through the magnetic field forces the electron to change its spin orientation such that it either aligns or anti-aligns with its z-component to be $\pm \hbar/2$. This is what distinguishes quantum physics from classical physics: the act of measuring a quantity will disturb the system.

(ii) The other components of the spin are indeterminate: if I were to pass these electrons into another magnetic field, say aligned with the x-axis, again it will be found that 50% of the electrons will align with the magnetic field (spin = $+\hbar/2$), and 50% will anti-align (spin = $-\hbar/2$), this time along the x-axis. On the other hand the spin along the z-axis is no longer known for these particles.

(iii) One way to mathematically represent this quantum system (read, the wave function) is this:

$$|\psi> = (1/2)^{\frac{1}{2}} (|\uparrow> - |\downarrow>) \qquad (2.7)$$

Now this is called a superposition of two quantum states, the up and down states. Note that if we want to calculate the probability that the electron has a spin up, we take the product of the vector $|\uparrow>$ with the wave function $|\psi>$, and square that.

$$P = ||<\uparrow|\psi>||^2 \qquad (2.8)$$

$$= 1/2 [<\uparrow|(|\uparrow> - |\downarrow>)]^2$$

$$= 1/2 [<\uparrow|\uparrow> - <\uparrow|\downarrow>]^2$$

Using the orthogonality condition, which is a fundamental property of a Hilbert space,

$$<\uparrow|\uparrow> = 1 \ and \ <\uparrow|\downarrow> = 0$$

We get,

$$P = 1/2, or \ 50\%, \qquad (2.9)$$

Which is what is observed in the lab.

(4) Now here comes the real crunch. Writing $|\psi> = (1/2)^{\frac{1}{2}} (|\uparrow> - |\downarrow>)$ is called a superposition but it's not meant to mean that the electron "lives" simultaneously in two states and can't make up its "mind" in which one it wants to live. Those states do not represent ordinary vectors of real objects - like velocities, acceleration, forces, as was mentioned above. If it were the case, then since these two vectors are equal in magnitude and opposite in direction I would be able to claim, $|\uparrow> = (-1)|\downarrow>$. And the orthogonality condition would no longer hold, and P would not equal to 50% - actually it would turn out to be 100%!!! What needs to be reminded is that the two vectors, $|\uparrow>$ and $|\downarrow>$ represent <u>possible states before the measurement</u> takes

place. And the beauty of it all is that they form a complete set of orthogonal unit vectors, in an abstract space called the Hilbert space, which provides a powerful method of calculating probabilities.

2.4 A Second Look at the Two-Slit Experiment

Two states will evolve, and we can write this process as,

$$|A> \rightarrow |A'>, \qquad |B> \rightarrow |B'>$$

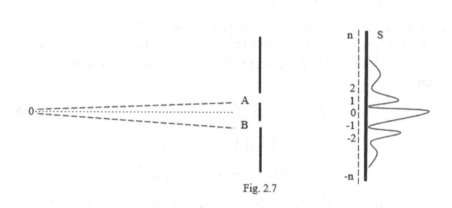

Fig. 2.7

By the principle of linear superposition, these two states will also evolve as,

$$|A> + |B> \rightarrow |A'> + |B'>$$

We label the position from which the electrons pass through as such: where they leave is 0; the slits are labelled A and B; and the screen, S (see fig. 2.7). The state representing the electron passing through slit A is denoted by $|A>$ and going through the second slit B as $|B>$. When the electrons leave position 0, from the symmetry of the setup, we can say that they arrive at position A and B with equal probability. So we write,

$$|0> \rightarrow |A> + |B>$$

When an electron has arrived at A or B, what happens after that when they hit the screen? Experiments show that they can land on any of the points on the screen, so we write,

28

$$| A > \; \rightarrow \; \Sigma_n \Psi_n \, | \, n >$$

$$| B > \; \rightarrow \; \Sigma_n \Phi_n \, | \, n >$$

Where $| \, n >$ forms a complete set of orthonormal vectors. The whole process can be described as,

$$| 0 > \rightarrow \; | A > \; + \; | B > \; = \; \Sigma_n (\Psi_n + \Phi_n) \, | \, n > \equiv \chi_n$$

The probability that an electron will arrive at the mth point on the screen is (equation 2.8),

$$P_m = \; \| < m \, | \, \chi_n > \|^2$$

$$= < \chi_n \, | \, m > < m \, | \, \chi_n >$$

$$= \Sigma_n \Sigma_{n'} (\Psi_{n'}^* + \Phi_{n'}^*) < n' \, | \, m > < m \, | \, n >$$
$$(\Psi_n + \Phi_n)$$

Using the orthogonality condition,

$$< n' \, | \, m > = \delta_{n'm} \; ; \; < m \, | \, n > = \delta_{nm}$$

$$P_m \; = \; (\Psi_m^* + \Phi_m^*) \, (\Psi_m + \Phi_m)$$

$$= \; \Psi_m^* \Psi_m + \Phi_m^* \Phi_m + \Psi_m^* \Phi_m + \Phi_m^* \Psi_m \quad (2.10)$$

The first term $\Psi_m^* \Psi_m$ represents the probability if only the first slit was open. Similarly, the second term $\Phi_m^* \Phi_m$ represents the probability if only the second slit was open. Classically, we should get the sum of these two terms if both slits were open. But we do not observe that. The interesting aspect of this result from quantum physics is that we get two extra terms, $\Psi_m^* \Phi_m$ and $\Phi_m^* \Psi_m$, that correctly explains the interference pattern of the double-slit experiment. Another major difference between classical physics and quantum physics is that in the first, probabilities are added, while in the second, the amplitudes are added first and then we square the amplitudes to get the probabilities.

2.5 The Act of Measuring

Suppose we want to know through which slit the electron has passed. This can be done by inserting a detector at position A. Furthermore, we prepare the electron at position 0 with a down spin. When it passes through A, its spin is flipped to an up spin, and when it passes through position B, nothing happens to the electron and it remains with a spin down (see fig. 2.8).

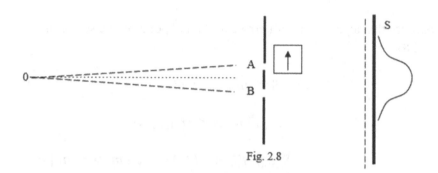

Fig. 2.8

We need two labels for the states: one for position, and the other for the spin. We describe the process as,

$$|0, d > \rightarrow \quad |A, u > \; + \; |B, d > \; \rightarrow \quad \Sigma_n \Psi_n \,|\, n, u > \; + \Sigma_n \Phi_n \,|\, n, d >$$

Due to the presence of the detector, the electrons are entangled through their spins: one is up, the other is down. Entanglement means that if we know a certain property of one particle, we also know the property of a second particle. In this case, we know that if the spin at A is up after passing through the detector, we also know that if it passed at B, it is spin down. Again to calculate the probability of finding the electron at the mth position, we square the amplitudes, or multiply the amplitude with their complex conjugate (equation 2.8).

$$P_m = (\Psi_m^* < m, u | \; + \; \Phi_m^* < m, d |) \, (\Psi_m \,|\, m, u > \; + \Phi_m | m, d >)$$

$$= \Psi_m^* \Psi_m < m, u \,|\, m, u > \quad + \; \Phi_m^* \Phi m < m, d \,|\, m, d >$$

$$+ \; \Psi_m^* \Phi_m \; < m, u \,|\, m, d > \; + \; \Phi_m^* \Psi_m \; < m, d \,|\, m, u >$$

30

Note that the up and down vector states, because the electrons have opposite spins, are now orthogonal to each other.

$$< m,u \mid m,d > \; = \; 0 \; = \; < m,d \mid m,u >$$

$$< m,u \mid m,u > \; = \; 1 \; = \; < m,d \mid m,d >$$

Therefore,

$$P_m = \; \Psi_m^* \, \Psi_m \; + \; \Phi_m^* \, \Phi_m \tag{2.11}$$

This result is completely different from equation 2.10. We see now that the very act of detecting the spin of one electron, that is, making some sort of measurement, destroys the interference pattern. Again, this is another markedly difference between a classical system, in which we can always make a measurement without disturbing it, and a quantum system, in which a measurement entails disturbing the system and getting a different result.

2.6 Bell's Theorem Revisited

In Bell's theorem[5], we make two assumptions in the proof. These are:

A. Logic is a valid way to reason.

B. Body either has a property A or doesn't have property A.

This is important to understand. The parameter in question is not necessarily non-locality. It can be anything that a particle possesses and can be measured. Consider the set of all measurements, for which A, B and C are any three measurements, and are independent property. Examples: A is up or down, B is head or tail, C is red or green, etc. Secondly, the theorem is not about hidden parameters but whether it has a property or not. Making it about non-local hidden parameters is to doubly compound the error in misinterpreting Bell's theorem.

Derivation of Bell's inequality

Definition: if an object has property A, we denote that as A+; if not, we denote it by A−

$$N(A+,B-) = N(A+,B-,C+) + N(A+,B-,C-) \tag{2.12}$$

This is true since an object must have the property C or does not have it.

$$N(A+, B-) \geq N(A+, B-, C-) \tag{2.13}$$

Since $N(A+, B-, C+)$ cannot be smaller than zero.

$$N(B+, C-) = N(A+, B+, C-) + N(A-, B+, C-) \tag{2.14}$$

Similar reasoning as above: an object must have the property A or does not have it.

$$N(B+, C-) \geq N(A+, B+, C-) \tag{2.15}$$

Similar reasoning as in equation 2.13: $N(B+, C-)$ cannot be smaller than zero.

Adding inequalities 2.13 and 2.15,

$$N(A+, B-) + N(B+, C-) \geq N(A+, B-, C-) + N(A+, B+, C-) \tag{2.16}$$

(6) But the RHS of 2.16 gives:

$$N(A+, B-, C-) + N(A+, B+, C-) = N(A+, C-) \tag{2.17}$$

That is, an object must have the property B or does not have it. Substituting 2.17 into 2.16, we get,

$$N(A+, B-) + N(B+, C-) \geq N(A+, C-) \tag{2.18}$$

And that completes the proof.

To reiterate: a body has a property which can be measured or it does not have that property. For instance, looking at the earth at a distance, one can observe its spin around an axis and measure it. On the other hand, looking at the moon, we observe it has no spin. So either a body has a spin (the earth) or it doesn't have a spin (the moon). On the other hand, the electron has a spin, but only one component can be measured. The other two components remain indeterminate once one component is measured as it was discussed through fig. 2.5 and 2.6. It is in that mind frame that we must interpret Bell's theorem. It goes without saying that a classical system will not violate Bell's theorem, while a quantum system will. In Alain Aspect experiment[6], Bell's inequality theorem was tested by measuring the polarization of photons along different axes. The inequality was violated as the system understudied was a quantum system. To attribute this violation to non-locality is a major blunder. The violation is strictly due to the HUP: along the three axes, the components of the spin are incompatible observables, and Bell's theorem applies only to a classical system. So applying Bell's theorem to a quantum system will inevitably result into a violation.

2.7 The EPR Revisited

It's time to go back to the 1927 Solvay Conference when the disagreement between Einstein and Bohr first surfaced and began the debate that has lasted ever since. The first disagreement centered on the notion of a wave collapse. Give Einstein one point (+1) being on the right side. He correctly deduced that such collapse would mean the existence of a spooky action at a distance.

The EPR[7] that came subsequently (1935) proposed that there were hidden parameters to explain what Einstein thought was the unexplainable.

Give Einstein a (- 1) point. And so it's a tie as far as Einstein is concerned.

Here's your typical argument that has come through the decades since this disagreement started.

A particle at O decays and sends two particles: an electron e⁻ towards Alice, and a positron e⁺ towards Bob (Fig. 2.9).

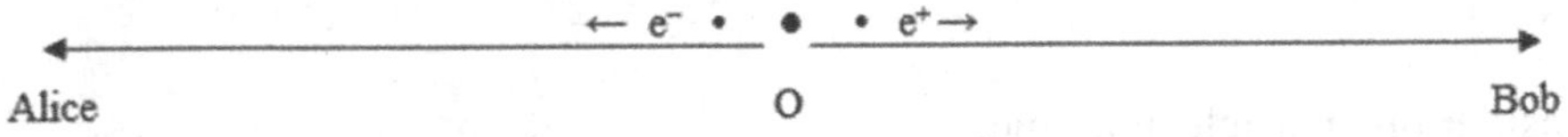

Fig. 2.9

Each particle flies off in opposite direction with opposite momentum (conservation of momentum) and opposite spin (conservation of angular momentum).

Case A

Alice is going to measure the spin of her particles with a magnetic field along the z-axis, likewise for Bob. So both are performing an experiment depicted in Fig 2.5-2.6. In each case, consider one particle at a time. As the electron (positron) approaches the magnetic field, the orientation of the spin with the magnetic field is totally unknown to our two observers. Now the particle goes through the magnetic field. There are only two possibilities for each particle,

(1A) Alice measures a spin up and Bob measures a spin down.

(2A) Alice measures a spin down and Bob measures a spin up.

There are NO other alternatives.

There is no mystery here, there is no spooky action at a distance, there is no weirdness, there is no communication traveling faster than light. There are only two possibilities, and this is what will be observed, which is explained entirely by the conservation laws.

On the whole, Alice will measure 50% of all the electrons coming her way with spin up, and 50% with spin down. Bob will measure similar results for his positrons.

Case B

Alice is going to measure the spin of her particles with a magnetic field along the z-axis, but this time, Bob will measure his particles along a different axis, say the x-axis.

The situation doesn't change in regard to the particle approaching the magnetic field: the orientation of the particle's spin is still unknown to both Alice and Bob.

Consider one particle at a time.

(1B) Alice measures the first particle with a spin up along the z-axis.

(2B) Bob measures his first particle with a spin up along the x-axis.

Can Bob conclude that he also knows that his particle has a spin down along the z-axis, since Alice measured her particle with a spin up along her z-axis?

No, he doesn't know. His experiment is different than Alice's as his particle's orientation was forced along the x-axis, by an amount that is unknown. And Alice's particle was forced to align along the z-axis by also an unknown quantity. The only conclusion that Bob can make is what he measures: a spin up along the x-axis. Secondly the components of the spin of his particle along the y and z axis remains unknown to him, just as Alice doesn't know the x and y components of her particle.

As in case A, Alice will measure 50% of all the electrons coming her way with spin up, and 50% with spin down, but keep in mind, she has only measured the spin along the z-axis. She has no knowledge of the other components of the spin of her particles – the x and y components.

Likewise Bob will also measure 50% of all the positrons coming his way with spin up, and 50% with spin down, but keep in mind, he has only measured the spin along the x-axis. He has no knowledge of the other components of the spin of his particles – the y and z components.

Again there is no mystery here, there is no spooky action at a distance, there is no weirdness, and there is no communication traveling faster than light.

2.8 Conclusion

How can we explain the confusion that has reigned for more than nine decades?

There were mistakes done at different levels:

(1) A misinterpretation of Bell's theorem in which the original intent did not include non-locality, but as a test to see whether or not a particle has a certain property that can be measured.
(2) A misinterpretation of the disagreement between Einstein and Bohr. Einstein's objection to the collapse of the wave function implied a spooky action at a distance, and Bohr should have listened to that.
(3) A misinterpretation that the wave function represents a real wave when in actuality it represents the possible states of a quantum system before a measurement.
(4) When Bell's theorem was violated by a quantum system, those violations were misinterpreted as evidence of an instantaneous collapse of the wave function and non-locality.

Those who were carrying the torch for Einstein thought that Bell's theorem confirmed non-locality (which Bell's theorem doesn't really say anything about non-locality nor hidden parameters) because that also confirmed that Einstein was right (a wave function collapse implied a spooky action at a distance, but the wave function isn't a real wave to begin with) leading to the idea that an instantaneous collapse (nothing can travel faster than the speed of light) makes the universe weird.

Here's the real deal: there is no instantaneous collapse of the wave function, and there is no spooky action at a distance.

Chapter 3

A New Kinematics – A Toy Model

Toy models play a vital role in physics. It is well-known that the de Sitter model in cosmology plays an important role in physical applications[8]. Three laws of kinematics are identified. The first two were already shown in section 1.4. The third law of kinematics is a new law of kinematics in a universe of bouncing balls in which there is no exchange of angular momentum (no spin involved), and it states: that for free particles, under no circumstances a body with higher kinetic energy can gain energy from another body with lower kinetic energy. Stated differently: in an elastic collision, it will always be the case that the higher kinetic energy body will always lose energy to the lower kinetic energy body, and the lower kinetic energy body will always gain energy from the higher kinetic energy body. Moreover, under general considerations from the theory of Special Relativity, a particle can decay into other particles only if there is a source of positive energy necessary to insure that the new particles have positive kinetic energy, which is why $E=mc^2$ provides the source of this positive energy in the form of mass conversion into energy. Also this new kinematics leads to the reformulation of the 2nd law of thermodynamics which explain why heat can only flow from a hot body to a cold body, and never the other way around, that a system left on its own will see its disorder increase in time, and that the entropy as defined according to Boltzmann tends to increase. The insight that we gain from this deeper concept which underlies the new law of kinematics and the reformulation of the second law of thermodynamics is that vacuum energy is a necessary condition for particles to pop out of the vacuum. However, this spells trouble for Hawking radiation as in the way it was formulated, it would lead to a violation of this new law of kinematics. Surprisingly this new law of kinematics also sheds new insights into Planck's ad hoc hypothesis, $E = \hbar\omega$, and why energy must be quantized.

3.1 First Law of Thermodynamics

The 1st law of thermodynamics states that the total energy of a system is always conserved. But it does not specify how the energy is redistributed after an elastic collision. Denote the higher kinetic energy particle as B-particle, B $\equiv$ "Big" and the lower kinetic energy particle as L-particle, L$\equiv$ "Little". The 1st law of thermodynamics states after an elastic collision,

$$KE_B + KE_L = KE'_B + KE'_L \tag{3.1}$$

Where KE $\equiv$ Kinetic Energy before collision, KE' $\equiv$ Kinetic Energy after collision.

3.2 The Second Law of Thermodynamics

The new law of kinematics states that,

$$KE_B > KE'_B \tag{3.2}$$

That is, the B-particle always loses energy after the elastic collision. Rewriting (3.1) as,

$$KE_B = KE'_B + KE'_L - KE_L \tag{3.3}$$

Substitute (3.3) into (3.2),

$$KE'_B + KE'_L - KE_L > KE'_B$$

$$KE'_L - KE_L > 0$$

$$KE'_L > KE_L \tag{3.4}$$

Consequently, after collision the corresponding L-particle always gains kinetic energy.

3.2.1 Arguments from Special Relativity

At the core of the theory of Special Relativity is that the laws of physics are the same in all inertial frames.

Case A

Before Collision (Fig. 3.1):

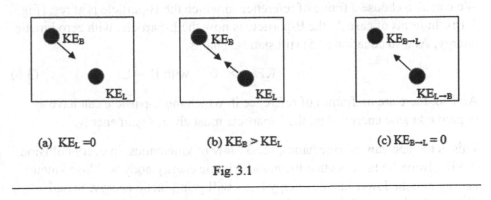

(a) $KE_L = 0$ (b) $KE_B > KE_L$ (c) $KE_{B \to L} = 0$

Fig. 3.1

In Fig. 3.1b, we have a case that could arise in the lab frame. We can always choose a different frame such that the L-particle is at rest, $KE_L = 0$, (Fig. 3.1a).

After Collision (Fig. 3.2)

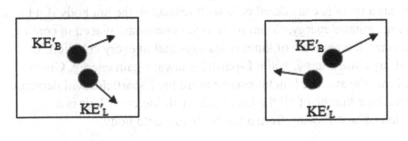

(a) $KE'_L > 0$ (b) $KE_B > KE'_B$

Fig. 3.2

We make the claim that a particle can never have negative kinetic energy[9]. From equation 3.4, after an elastic collision, the lower kinetic energy will then have,

$$KE'_L > 0 \qquad (3.5)$$

And therefore the L-particle must gain energy.

Case B

39

We can also choose a frame of reference in which the B-particle is at rest (Fig. 3.1c). In terms of case A, the B-particle is now the L-particle, with zero kinetic energy. And so equation (3.5) still stands. That is,

$$KE'_L > 0 \quad \text{with B} \rightarrow L \tag{3.6}$$

And so, there are no frames of reference in which the L-particle can have negative kinetic energy. And the L-particle must always gain energy.

This is the new law of kinematics, the 3rd law of kinematics: in every collision, it will always be the case that the higher kinetic energy body will lose kinetic energy and the lower kinetic energy body will gain kinetic energy. Stated differently: in every frame of reference, the B-particle loses kinetic energy, and the L-particle gains kinetic energy. There are no exceptions.

In what ways this leads to a reformulation of the 2nd law of Thermodynamics?

3.3 Heat Transfer from Hot to Cold

A hot body is a system in which the average kinetic energy of its constituents is high, whereas a body is considered cold with respect to the hot body if it has a lower average kinetic energy. When these two systems are placed in contact with each other, the new law of kinematics says that in every collision, B-particles always lose energy, while L-particles always gain energy. Given sufficient time, the average kinetic energy of all the B-particles will decrease, while the average kinetic of all the L-particles will increase. This is a manifestation of heat flowing from a hot body to a cold body.

Let us demonstrate why this is so.

Consider that for the hot body, the system is made up of B-particles. For every collision taking place when the two system are in contact, we get the following,

$$KE_{B1} > KE'_{B1}$$

$$KE_{B2} > KE'_{B2}$$

$$KE_{B3} > KE'_{B3}$$

$$....$$

$$KE_{Bi} > KE'_{Bi}$$

We get,

$$\sum_i KE_{Bi} > \sum_i KE'_{Bi} \qquad (3.7)$$

The last step was obtained by summing up over all the B-particles. But we're missing one element to work out the average kinetic energy. The temperature is a measure of the average kinetic energy, and so we need to divide the sum by the number of particles. Once the two system are in contact, the question is, will the number of B-particles remains constant, or will it change? To answer this, we will use the following illustration.

We will consider a system of three particles where $KE_1 > KE_2 > KE_3$, (Table 3.1)

Case	Collision	KE$_1$	KE$_2$	KE$_3$	N$_B$	N$_L$	Total Energy	Ave. Total Energy
1	none	30	20	10	1	1	60	20
2	1↔2	29	21	10	1	1	60	20
3	2↔3	29	20	11	1	1	60	20
4	1↔3	27	20	13	1	1	60	20
5	1↔2	26	21	13	1	1	60	20
6	2↔3	26	19	15	1	1	60	20
7	1↔3	24	19	17	1	1	60	20
8	1↔2	23	20	17	1	1	60	20
9	2↔3	23	19	18	1	1	60	20
10	1↔3	21	19	20	1	1	60	20

Table 3.1

(a) In case 1, we have no collision. This is our starting point. The energy assigned to the particles are arbitrary and will be sufficient to illustrate what is happening just by applying the new law of kinematics. So we have one B-particle, particle 1 (N$_B$= 1), and one L-particle, particle 3 (N$_L$= 1). We ignore particle 2 for reasons that will be obvious.

(b) We notice that particle 2 with a kinetic energy close to the average kinetic energy (Total energy = 60, Ave. KE = 20) fluctuates around the average. When it interacts with particle 1, it is the L- particle, and so it

41

gains energy (cases 2, 5, 8). When it interacts with particle 3, it is the B-particle, so it loses energy (cases 3, 6, 9). So we don't count it in the N_B column nor in the N_L column.

(c) Particle 1 is always the B-particle, so it loses energy in every collision.
(d) Particle 3 is always the L-particle, so it gains energy (except for case 10, which we will discuss later). The net result is that for all particles, their kinetic energy tends towards the average kinetic energy.
(e) In case 10, the last entry, particle 2 has less energy than particle 3. We just need to relabel, $2 \rightarrow 3$, and $3 \rightarrow 2$. And so $N_B = 1$ and $N_L = 1$.

In light of the reformulation of the 2nd law, we can look at our B-particles, in the system we've designated as the hot body, as particles wearing a tag which reads "B". Similarly, for the cold system, the particles are wearing tags with the label, "L". So now we let them loose in the same room. They start bumping into each other. Suppose in that process that one of the L-particles has gained enough energy so that it has earned to be part of the B-team. Call it the "lucky" particle. All we need to do is switch tags: there is at least one B-particle that has less energy than lucky particle, otherwise our lucky particle isn't "lucky". After the switch, the number of B-particles and L-particles, both remain the same. We can do that for every single collision: either we switch tags or we don't.

We can safely say that the number of B-particles (N_B) remains constant, as well as for the L-particles (N_L). So to calculate the average kinetic energy, we add the energy of each particle divided by the number of particles. Equation 3.7 becomes,

$$\sum_i KE_{Bi} / N_B \; > \; \sum_i KE'_{Bi} \, / \, N_B \qquad (3.8)$$

Or

$$Ave. KE_B \; > \; Ave. KE'_B \qquad (3.9)$$

And this can only hold if N_B is the same throughout the process when the two systems are placed in contact, which our table demonstrates. Recall that the average kinetic energy is also a measure of the temperature.

3.4 Thermal Equilibrium

In standard interpretation of thermodynamics, equilibrium is defined in terms of an isolated system which is found with equal probability in each one of its accessible states [10]. In our case, we redefine thermal equilibrium as a limiting process. We need to examine table 1 and compare each particle's kinetic energy with the average kinetic energy of the combined system. We notice that the difference of the B-particle's kinetic energy with the total system's average energy tends to decrease towards the average kinetic energy of the combined system. We can express that as,

$$\text{Ave. } KE_B \rightarrow \text{Ave. } KE \text{ of the combined system,} \qquad (3.10)$$

And also, the difference of the L-particle's kinetic energy with the total system's average energy tends to increase to the same average.

$$\text{Ave. } KE_L \rightarrow \text{Ave. } KE \text{ of the combined system.} \qquad (3.11)$$

Since temperature is a measure of the average kinetic energy for any given system, for the combined system in our case, this temperature is the equilibrium towards which the kinetic energy of both the B-particles and the L-particles are moving. The result is that the number of B-particles (and the L-particles) remains constant. When the combined system has finally reached thermal equilibrium, the B-particle (the L-particles) being near the equilibrium will lose (gain) kinetic energy sufficiently, requiring a frequent change of tags, nevertheless, the collisions even as tiny these differences in kinetic energy can get will continue relentlessly, which can be observed as fluctuations. Thus, a thermometer sensitive enough will be able to show those fluctuations.

The combination of the 1st and 2nd laws makes this process necessary, and not just probable. This is a manifestation that heat flows from a hot body to a cold body such that the combined system will eventually reach thermal equilibrium.

3.5 Positive Definite

So the question arises, how does this reformulation fit in the scheme of Boltzmann's microscopic states? Boltzmann had to assume that all those states were equally probable, and so it necessitated the use of probability theory in which each outcome is a number between 0 and 1, and the sum of all outcomes is 1. What this means is that Probability theory uses the concept of positive definite – you can't have negative probabilities. It turns out that kinetic energy is also positive definite, [see equation (3.5)]. However he assumed that the energy distribution could take values that we see would be wrong according to the new law of kinematics. If we look again at table 3.1, Boltzmann assumed that the B-particles could equally take any value between 0 and 60; similarly with the L-particle, any value with equal probability between 0 and 60 as long as the law of conservation of energy is obeyed. What the new law of kinematics says is that the B-particle in that particular frame of reference can take values with equal probability but only between 30 and 20, while the L-particles can only take those values with equal probability only between 10 and 20.

Needless to say that in a different frame of reference these ranges of values would change. In case A above, the L-particle being at rest would take values of energy ranging between 0 and 20 after collision, while the B-particle would be in the range 20 to 40 in the aftermath of the collision. In case B, those numbers would be reversed. So we can see that the number of microscopic states available to each particles is restricted and depends on the frame of reference.

Though Boltzmann's work was a step in the right direction, his analysis of ensemble of particles obscured what was happening at the individual level, where one particle collides with another particle. And only by taking into account this new law of kinematics and a reformulation of the 2nd law can bring to the fore why heat always flow from hot to cold.

We can now claim that we've inherited all the mechanism of what has been established in thermodynamics.

We are all familiar with an ice cube left in the open, and finding out later on it has evaporated. The standard description is that the atoms in the cube gain energy from the air molecules sufficient enough to escape their bondage in the cubic crystal. This is often interpreted as the entropy increases, that is, the atoms arranged in the cube in a certain order but later on, as free particles in the air, they have less order, or greater disorder. Often, the belief is that some molecules in the air could join the water molecules in the cube, but that the probability for that to happen is so small, so insignificant that we can ignore that. We are also familiar with the scenario often repeated that the air in the room could all gather in one corner, leaving everyone in the room gasping for air. But have no fear we are told, not to worry that for this to happen, it is very unlikely.

With the reformulation of the 2nd law we can say that the air molecules are the "Big" particles in the room, those in the ice cube are the "Little" particles. The average kinetic energy of the "Big" particle tends to decrease to some equilibrium. Unless the outside temperature is well below zero centigrade, which is the temperature at equilibrium for this case, the air molecules in their tendency to lose kinetic energy up to thermal equilibrium will still have too much energy to be solidified as the water molecules are. So air molecules freezing to join the ice cube is not going to happen.

In the case of the air molecules all moving towards the corner of the room, we would observe an increase in temperature in that corner, and for that to happen, the "Big" particles which have kinetic energies slightly above thermal equilibrium would have to gain energy, and we know this is forbidden. Unless there is a force shooing in these particles into one single corner, no such phenomenon is going to happen.

3.6 $E = mc^2$ Revisited

Consider a particle at rest that decays into two particles,

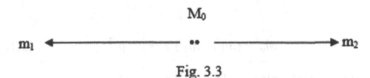

$$M_0$$

$$m_1 \longleftarrow \bullet\bullet \longrightarrow m_2$$

Fig. 3.3

Initially, the total energy is zero. After it decays into two particles, they will fly away from each other with equal and opposite momentum (conservation of momentum) and with equal and opposite spin (conservation of angular momentum). But they both have kinetic energy, and by the conservation of energy,

$$0 = KE_1 + KE_2 \tag{3.12}$$

This necessitates that one of the particles would carry NEGATIVE kinetic energy?! However according to the new law of kinematics, this is forbidden. The only way this can happen is that some positive energy is given to these two particles, and Einstein provided the answer in $E = mc^2$, that is, there is a mechanism by which this process can take place. I will make this statement stronger: It would be impossible for decay to occur in nature if mass could not be converted to energy. What Einstein discovered, although admittedly by other means, is that the existing laws of kinematics had to be amended in view of the constancy not only of the speed of light but all laws of physics in all inertial frames by making everyone aware that the Galilean transformations must be replaced by the Lorentz transformations – an idea that was successfully incorporated in Quantum Mechanics to yield Quantum Field Theory. Einstein revised the above equation to read as,

$$(\text{Mass} \rightarrow \text{energy}) = KE_1 + KE_2 \tag{3.13}$$

More specifically, the above equation is worked out to give for one particle (c=1),

$$E^2 = p^2 + m^2 \tag{3.14}$$

In the case of a particle of mass M_0 decaying into two particles of mass m_1 and m_2, we get

$$M_0^2 = p_1^2 + m_1^2 + p_2^2 + m_2^2 \tag{3.15}$$

Rearranging,

$$M_0^2 - (m_1^2 + m_2^2) = p_1^2 + p_2^2 \tag{3.16}$$

The condition for decay to take place and making sure it obeys the new kinematics, that is, no particle can have negative energy in any frame of reference is,

$$M_0^2 > (m_1^2 + m_2^2) \tag{3.17}$$

The mass can change as long as it obeys this restriction. The implication is a mass can decay through different channels, an observation that has been confirmed multiple times in high energy physics. However a re-interpretation is necessary in light of the new kinematics: whenever new particles appear, there must be a source of energy such that no new particle can have negative kinetic energy.

3.7 Vacuum Energy

It was Schrödinger who came up with the Klein-Gordon equation in his search to describe de Broglie waves. But it gave out negative frequencies and negative probabilities. Hence he abandoned it and used a non-relativistic approximation in his work. Dirac was able to give an explanation in terms of a negative sea with negative particles and the condition that it was all filled up. Occasionally a hole would respond to electric fields as though it were a positively charged particle and predicted the existence of anti-matter, which was discovered subsequently a few years later. Today we treat the positron as a "real" particle rather than a hole or the absence of a particle, and the vacuum is thought as the state in which no particles exist instead of an infinite sea of particles. On the other hand quantum field theory postulates that at every point in space, there is a field made of an infinite number of harmonic oscillators, which are expressed as Fourier series and annihilation and creation operators. With what has been established in this chapter, we can say that it is only through this device of a zero-point energy that quantum fluctuations are a possibility. So even though some are uncomfortable with this infinite zero-point energy, it is a necessity as it is a testimony that the new law of kinematics must prevail: no particles can pop out of the vacuum without a source to supply it with energy sufficient enough so that these particles can escape with positive kinetic energy. What remains to be determined is how big this ground zero energy is as it is at variance with Dark Energy by the most mismatched ratio in science history. Hopefully the new law of kinematics and this reformulation of the 2nd law of thermodynamics are steps in that direction.

3.8 Hawking Radiation

Hawking[11][12] assumed that particles can pop out of the vacuum. As it was pointed out above, this process is permissible as long as we take the vacuum energy as necessary to allow this process. And this is further emphasized in the Casimir force and the Lamb shift, to name two instances. In the case of Hawking radiation, it was argued that the Black Hole loses energy by absorbing one of the free particle with negative kinetic energy, and the other free particle just outside the Horizon escapes to infinity [13]. In the process, the Black Hole's mass would decrease, and Hawking formulated that its mass would be inversely proportional to its temperature. This is all well within the 1st law of thermodynamics. What is no longer acceptable in the context of this new law of kinematics is that one of the particles has negative kinetic energy.

So unless there is an unknown mechanism to provide positive kinetic energy, the process cannot take place. And if there is such a mechanism, that particle would have positive kinetic energy making the argument that the Black Hole loses mass untenable. Notwithstanding this puts a hole in the Information Paradox[14] and the Firewall Controversy[15].

3.9 Center of Kinetic Energy

Definition: The center of kinetic energy is that frame of reference in which both particles, the B-particle and the L-particle, would have their kinetic energy equal to the average kinetic energy. Note: The center of kinetic energy is an idealized frame of reference. It cannot be realized in the real world as one would have to measure the kinetic energy to an infinite precision.

Definition: The deviation is the energy difference between the body's kinetic energy in a given frame and its kinetic energy in the frame of the center of kinetic energy. Irrespective of how we define our energy scales according to one frame of reference, what matters is that we measure energy differences. And this difference, or the deviation, is frame invariant [16].

So the questions arises: consider the B-particle, which we claim that it will always lose energy in a given collision, could it loses all of its energy to the L-particle?

BEFORE COLLISION

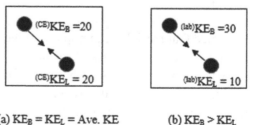

(a) $KE_B = KE_L = $ Ave. KE   (b) $KE_B > KE_L$

Fig. 3.4

Going back to table 1, where case 1 is the starting point with $^{(lab)}KE_B = 30$ and $^{(lab)}KE_L = 10$, (Fig. 3.4b); while Fig. 3.4a depicts the situation for an observer in the frame of reference which is the center of kinetic energy. For this case, we have $^{(CE)}KE_B = 20$ and $^{(CE)}KE_L = 20$. In terms of the deviation we write,

$$^{(lab)}KE_B - {}^{(CE)}KE_B = \text{Dev. KE} \tag{3.18}$$

$$^{(lab)}KE_L - {}^{(CE)}KE_L = -\text{Dev. KE} \tag{3.19}$$

Using the values for fig 3.3,

$$\text{Dev. KE} = 10 \tag{3.20}$$

Suppose that the B-particle loses all of its energy to the L-particle. Then what we would see after collision, (after applying equations 3.18, 3.19 and 3.20) is,

AFTER COLLISION (hypothetically)

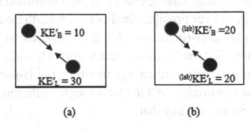

(a)   (b)

Fig. 3.5

After Collision (considering thermal equilibrium as the limiting process)

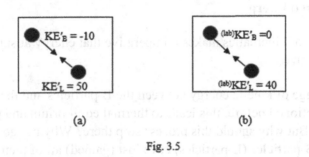

(a)   (b)

Fig. 3.6

Note that the average kinetic energy hasn't changed, but Fig. 3.5a no longer depicts the center of energy, and KE'_B has now negative kinetic energy in that frame, and this is forbidden. And so this is an unphysical situation.

Consider Fig. 3.6:

(i) Note that the lab frame is now the center of energy as the B-party (L-particle) has lost (gained) the maximum to bring it to the average kinetic of the system, and no particle has negative kinetic energy.

(ii) Note that the center of energy serves as a litmus check if the 3rd law of kinematics was correctly applied.

And so, we can conclude that the B-particle will lose energy in the lab frame as depicted in table 3.1, but will never lose all of its energy. The maximum it can lose is up to it reaches the average kinetic energy at the thermal equilibrium.

3.10 Quantized Energy

The third law of kinematics makes it imperative that energy at sub-microscopic scales is quantized.

In the exchange of kinetic energy between the B-particles and the L-particles, as it was mentioned before, this leads to thermal equilibrium and produces fluctuations. But why should this process stop there? Why not go all the way until all the B-particles (L-particles) have lost (gained) all of their energy permissible, in which their kinetic energy would then equal exactly the average kinetic energy as depicted in Fig 3.5b? If that would happen, then all of the B-particles and all the L-particles are identical, in that, they are all have equal in kinetic energy, the average kinetic energy of the combined system, and no exchange of energy is ever possible. In this universe, there are only bouncing balls, in which all the B-particles and all the L-particles are the same. Call that the heat catastrophe. But that's not the real world. The only way out is that there exists a minimum non-zero energy in which the B-particles can still exchange with the L-particles, and in every single collision we either exchange tags or we don't. So we can say that,

$$E_{min} = k(?) \tag{3.21}$$

Where the k is a constant that represents some minimum value, and the question mark (?) represents a quantity of which the energy can be a function.

And so what is/are the candidate(s) possible suitable for that unknown quantity?

3.11 The Harmonic Oscillator

Quantum Field Theory (QFT) began with the works of Dirac and others, and the key idea was to unify Quantum Mechanics (QM) with SR. By demanding that the Lagrangian is Lorentz invariant (the 2nd law of kinematics), it guarantees that SR is incorporated into the theory. From QM, what was used essentially is the quantized harmonic oscillator – the only problem with an exact solution to the Schrödinger's equation – which contains everything you need in QM.

In our previous assumptions, it was stated that no energy enters our system, no energy leaves our system, and we just have free particles. It was also stated that if a particle breaks up into two or more particles, a source of energy had to be available to give each particle positive kinetic energy, and that was the mass in the form of $E = mc^2$. In an elastic collision what we're dealing with is not a break-up, but just an exchange of energy. So mass as a possible candidate for our unknown quantity is ruled out. Consider now that our particles are really harmonic oscillators, the only remaining candidate would be their frequencies, and that's what oscillators do – they oscillate at a certain frequency. So we can write equation 3.21 as,

$$E_{min} = k\omega \qquad (3.22)$$

This would be the minimum quantity of energy that must be exchanged in order to avoid the heat catastrophe. But what still remains to be shown is that the energy is quantized.

Let us assume that in any elastic collision, the energy exchanged can be expressed as,

$$E_{exchanged} = nE_{min} = nk\omega \qquad (3.23)$$

All we need to do is to show that n is an integer.

3.12 Normal Modes

And now we go back to Planck and the experiment he needed to explain. What was used in those days is that radiation inside a black box was made up of frequencies of all modes. That was permissible as Fourier had successfully demonstrated that such frequencies can expressed as a series of fundamental modes, also called normal modes. More specifically, we will be concerned with the wavelength,

$$c = \lambda v \qquad (3.24)$$

Multiply both sides by 2π, and $\omega = 2\pi v$, rearranging,

$$\omega = 2\pi c/\lambda \qquad (3.25)$$

But these normal modes obey boundary conditions in the box, such that the 1st mode is $L = \frac{1}{2} \lambda$, the 2nd mode is, $L = \lambda$, the third, is $L = 1\frac{1}{2} \lambda \dots L = (n + \frac{1}{2}) \lambda$, where n is an integer. Substitute that into equation 3.25,

$$\omega_{quantized} = (2\pi c/L)(n + \frac{1}{2}) \qquad (3.26)$$

With this result, we can repackage equation 3.23 as,

$$E_{exchanged} = nk\omega = (n + \frac{1}{2})\hbar\omega \qquad (3.27)$$

Where $\hbar$ is a universal constant known as the Planck constant (in some text, the reduced Planck constant), and the energy is quantized.

3.13 Quantum Field Theory

This part is more of a suggestive nature. We can point to where the new kinematics could be playing a role unsuspectedly in QFT. The φ^4-theory was particularly chosen for illustration purposes.

One of the major problems encountered in the development of QFT that had to be overcome was the many infinities that plagued the theory from its onset. This necessitated several methods to remove these infinities that resulted from different considerations. One of these methods is known under the name of adding counter terms. So for a scalar field, you start with a Lagrangian that comes from the Klein-Gordon equation – an equation which is the quantized version of equation (3.14) and to that, there is an interaction that is added [17],

$$\mathcal{L} = \tfrac{1}{2}(\partial_\mu\varphi)^2 - \tfrac{1}{2}m^2\varphi^2 - \tfrac{1}{4!}\lambda\,\varphi^4 \qquad (3.28)$$

Under the scheme of renormalization, which we will not be exploring in this paper, the counter terms are added in order to subtract the infinities, and so the above equation is modified as

$$\mathcal{L} = \tfrac{1}{2}(\partial_\mu\varphi_r)^2 - \tfrac{1}{2}m^2\varphi_r^2 - \tfrac{1}{4!}\lambda\,\varphi_r^4$$

$$- \tfrac{1}{2}B(\partial_\mu\varphi_r)^2 + \tfrac{1}{2}A\,\varphi_r^2 + \tfrac{1}{4!}C\,\varphi_r^4 \qquad (3.29)$$

We see that the kinetic term is renormalized as,

$$\tfrac{1}{2}(\partial_\mu\varphi)^2 \to \tfrac{1}{2}(\partial_\mu\varphi_r)^2 - \tfrac{1}{2}B(\partial_\mu\varphi_r)^2 \qquad (3.30)$$

Similarly for the mass term,

$$\tfrac{1}{2}m^2\varphi^2 \to \tfrac{1}{2}m^2\varphi_r^2 - \tfrac{1}{2}A\,\varphi_r^2 \qquad (3.31)$$

The interaction term is also modified,

$$\tfrac{1}{4!}\lambda\,\varphi^4 \to \tfrac{1}{4!}\lambda\,\varphi_r^4 - \tfrac{1}{4!}C\,\varphi_r^4 \qquad (3.32)$$

In light of what has been established in this chapter, we can see that the K-G equation in its original form was not taking into consideration the new kinematics and had to be modified, in particularly (a) the kinetic term – the distribution of kinetic energy among new particles have restrictions as in equation (3.2); (b) the mass term – as indicated with equation 3.17 that mass though it must obey this restriction can assume different values when it converts to energy ; and consequently (3) this new kinematics also affects the interaction (the dynamics) of the theory. The 3rd law of kinematics has the potential of shedding new light on the origin of these infinities.

3.14 Conclusion

In this toy model we find a new law of kinematics. Stated again: it will always be the case that the higher kinetic energy particle will lose kinetic energy in an elastic collision, and the lower kinetic energy particle will gain energy. As a corollary, a free particle can never have negative kinetic energy. This gives new perspectives to the 2nd law of thermodynamics which is reformulated as: when two systems at different temperatures are place in contact with each other, the average kinetic energy of the higher kinetic energetic particles will decrease to the limiting point called the thermal equilibrium, while the average kinetic energy of the lower kinetic energetic particles will increase to the thermal equilibrium. This new law of kinematics also explains that decay of particles are possible only because a source is available through Einstein's $E = mc^2$, to give each particle after decay positive kinetic energy. It also provides an explanation why the zero-point energy is a necessary condition for the possibility of particles to pop out of the vacuum.

References

[1]Nima Arkani-Hamed; Messenger Lectures on "The Future of Fundamental Physics" Oct. 4, 2010.

[2] Steven Weinberg; Gravitation and Cosmology, John Wiley & Sons, 1972, pages 3, 19.

 [3]James B. Hartle; Gravity, an Introduction to Einstein's General Relativity, Addison-Wesley, 2003, page 189.

[4]Ta-Pei Cheng; Relativity, Gravitation and Cosmology, Oxford University Press, 2012, page 108.

[5] J.S. Bell (1964); "On the Einstein-Podolsky-Rosen Paradox", Physics 1: 195–200.

[6]Alain Aspect, Philippe Grangier, Gérard Roger (1982); "Experimental Realization of Einstein-Podolsky-Rosen-Bohm Gedankenexperiment: A New Violation of Bell's Inequalities", Phys. Rev. Lett. 49 (2): 91–4.

[7] A. Einstein, B. Podolsky, N Rosen; Physical Review 47(10) 777-780, bicode 1935 PhRv47.77E.

[8]V. Mukhanov; Physical Foundations of Cosmology, Cambridge University Press, 2013, page 35.

[9] Matthew D. Schwartz; Quantum Field Theory, Cambridge University Press, 2015, page 215.

[10] F. Reif; Statistical Physics, Berkeley Physics Course – Vol 5, McGraw-Hill Book Company, 1967, page 115.

[11] S. Hawking; Nature 248, 30 (1974).

[12] S. Hawking; Comm. Math. Phys 43, 199 (1975).

[13] Ta-Pei Cheng; Relativity, Gravitation and Cosmology, Oxford University Press, 2012, page 159.

[14]Leonard Susskind, The Black Hole War, Back Bay Company, 2008. Pp209-210.

[15] Almheiri, Ahmed; Marolf, Donald; Polchinski, Joseph; Sully, James (11 February 2013). "Black holes: complementarity or firewalls?" Journal of High Energy Physics **2013** (2).

[16] DOES THE INERTIA OF A BODY DEPEND UPON ITS ENERGY-CONTENT? By A. EINSTEIN, September 27, 1905.

English translation available at:
http://www.fourmilab.ch/etexts/einstein/E_mc2/e_mc2.pdf .

[17] T. Lancaster, S.T. Blundell; Quantum Field Theory for the Gifted Amateur, Oxford University Press, 2014, page 291.

Printed in the United States
By Bookmasters